Elon Musk

I0463439

Elon Musk's Best Lessons forever, Business, Success and Entrepreneurship!

Miguel Lopez

©Copyright 2017

Disclaimer

The information provided in this book is designed to provide helpful information on the subjects discussed. The author's books are only meant to provide the reader with the basics knowledge of the topic in question, without any warranties regarding whether the reader will, or will not, be able to incorporate and apply all the information provided. Although the writer will make his best effort share her insights, the topic in question is a complex one, and each person needs a different timeframe to fully incorporate new information. Neither this book, nor any of the author's books constitute a promise that the reader will learn anything within a certain timeframe.

Table of Contents   Page

Introduction

Much obliged and congrats on grabbing this book - Elon Musk: Best Lessons forever, Business, Success and Entrepreneurship!

This is the as of late redesigned second version of this book, finish with a few extra sections and upgrades! At the point when attempting to enhance your own present situation it is dependably an incredible thought to take a gander at the effective propensities for the individuals who preceded and there are few better cases of accomplishment in the twenty-first century than Elon Musk.

More than only a smart specialist, designer or creator, Musk is really a visionary, somebody who is attempting to enhance the world not on the grounds that it can profit, or even in light of the fact that it is the correct thing to do, however just in light of the fact that he sees mankind's actual potential and needs it to stick around sufficiently long to happen as expected.

Inside you will discover a record of Musk's life from his time as a youthful casualty of harassing in South Africa to making his first business, to being the CEO of a couple of organizations worth more than a billion dollars each. There is bounty to gain from Musk's life and a top to bottom talk of those lessons can be found in the last section. A debt of gratitude is in order for acquiring this book, I trust it gives you all that you are searching for. Appreciate!

Chapter 1: Become acquainted with Elon Musk

While Elon is the primary Musk that the general population of the twenty first century have known about, it isn't the first occasion when one of their positions has ascended to recognized unmistakable quality on the universal stage. Elon's grandparents were the first to go via plane from Australia to Africa, his extraordinary granddad won an all around advanced race over the length of Africa and his other grandma was the principal lady to ever hold a Canadian chiropractor's permit.

With an ancestry like this, the Musk family has dependably considered itself pioneers, a connection that the most recent, and perhaps most prominent, Musk has acknowledged all through his grown-up life. As such, he has been included in making a standout amongst the most omnipresent sites all through the entire of the web, gotten behind electric autos altogether and set his sights on interplanetary space go inside the following 20 years - and all before his fiftieth birthday!

Musk is additionally an image of the American dream, regardless of not being American, he is genuinely independent, working his way through school and beginning his first business so as to exploit the blossoming web scene to wind up distinctly a tycoon before the age of 30. Dissimilar to many individuals who may surmise that was sufficient, Musk rather put his cash into another startup which quintupled his total assets when his organization was obtained by eBay.

Still not fulfilled, Musk then sunk the sum of what he had earned up until that point, into wagering on the future and wagering huge. This is the thing that at last lead him to the standard spotlight as he went up against a main part at Tesla Motors and SpaceX. He is additionally vigorously put resources into sun powered vitality, owning a noteworthy share of one of the greatest sun oriented power makers in California.

He is additionally effectively financing what is likely the eventual fate of long separation head out by making what is known as the Hyperloop a reality.

At long last, when he finds a touch of extra time, he is attempting to guarantee that what he feels is the unavoidable making of computerized reasoning doesn't wind up working out in any capacity that doesn't profit mankind all in all. What it comes down to what is a modern and farfetched sounding innovation that will improve the world a place down the line, then it is an easy win that Musk is on the bleeding edge and has officially dedicated more than a million dollars to guarantee it turns into a reality.

While surely of virtuoso level knowledge, that isn't what makes Musk such a positive hotspot for change; no, that respect goes to his capacity to foresee likely future results in view of present pointers and his capacity rapidly assimilate endless measures of new data and use it appropriately.

This capacity can be seen most as of late through his SpaceX wander as when he got to be distinctly keen on really sending something to Mars he began by perusing all that he could about advanced science and conversing with a bundle of scientific geniuses. Starting there on he could precisely foresee where the market for such things was going on the grounds that he exhaustively comprehended the science behind it.

With his one of a kind mix of learning procurement abilities, combined with a business astuteness that has been demonstrated on numerous occasions, Musk could, without question, do pretty much anything he needs. Humankind all in all is to a great degree blessed that what he has settled on is making the world into the sci-fi perfect world researchers have been pursuing for a hundred years.

Chapter 2: Growing Up

The man who many individuals are calling the cutting edge Henry Ford was destined to Maye and Errol Musk on June 28, 1971 in South Africa. The eldest of three youngsters, Elon rapidly took after his dad, an electrical architect, keeping in mind frequently viewed as independent and calm, by the age of 10 he was at that point programing in BASIC.

He showed himself this expertise on his dad's Commodore VIC-20 and immediately utilized the abilities he figured out how to begin his first business. By the age of 12 he had programed his own particular amusement, titled Blastar, and rapidly sold it to a neighborhood magazine organization who possessed a distribution devoted to the developing marvel for an aggregate benefit of $500.

In spite of his initial entrepreneurial achievement, youth was often troublesome for the youthful Musk who was little for his age and was much of the time tormented by the other kids at the various tuition based schools he went to. This harassing developed to be severe to the point that while in center school he was really tossed down a flight of stairs and beaten so gravely that he was hospitalized after the assault.

When Musk was an adolescent, the point of Apartheid was fervently and he was searching for any approach to keep away from his obligatory administration in the South African military who busied themselves amid this period by effectively quieting those standing in opposition to the practice. At 17 years old, and completed with optional school, he initially attempted to access the United States and its prospering PC innovation scene however was denied section into the nation.

Not one to be hindered by an underlying refusal, he rather went to Canada in 1989 utilizing his mom's Canadian legacy to increase simple passage into the nation. School   With his time in the military stayed away from, Musk had done little to really enhance his parcel and he spent the following year living hand to mouth, sparing each penny he could so as to bear the cost of educational cost at a nearby school.

He wasn't fussy about the sort of work he did either, buckling down, humble employments tending vegetables, scooping grain, slashing wood and notwithstanding cleaning the kettle at a nearby observed factory, an assignment that few stayed with for any timeframe. When he was told this at the time, Musk wasn't certain what all the whine was about; he discovered rapidly, be that as it may, as the employment required a hazardous materials suit and for him to remain in a confined, hot and risky space for a few hours on end.

In spite of the trouble, Musk drove forward and endured essentially longer than most other individuals who endeavored the undertaking. With his school educational cost paid for, Musk could gain a spot at the Ontario-based Queens University. He put in 2 years going to the organization and investing energy with his mom and more youthful sibling Kimbal.

In their extra time, he and Kimbal would read the daily paper and after that call the intriguing individuals they read about and attempt and eat with them. One such individual they called was the president of a nearby bank, the subsequent lunch landed Musk a temporary job and a staunch supporter in his future attempts.

Now Musk was only 18, however he was at that point looking towards the future and how he could make his blemish on it. To such an extent, that the bank administrator's little girl can in any case recall a discussion they had at Musk's eighteenth birthday. It was about electric autos.

By 1992, Musk had separate himself from his associates by exceeding expectations at his reviews, to such an extent that he could gain a grant to a school in the United States. He spent the following year learning at the University of Pennsylvania before winning a four year certification in Physics before procuring another the next year in Economics.

Notwithstanding his obvious achievement, Musk wound up engaging against despondency amid this period. He swung to the religious writings of different religions for direction, before at last finding what he was searching for in a book by Douglas Adams called The Hitchhiker's Guide to the Galaxy.

While perusing the book, he went over a section which uncovered to the primary character that the significance of life was 42. Where many individuals would see a straightforward joke, this answer stayed with Musk and showed him the significance of asking the right question in any circumstance. With the response to the importance of life ringing in his brain, he contemplated it for some time and after that solicited himself what he thought from as the correct question.

This question was what advances were likely going to have the best impact on the whole human race sooner rather than later. His answers were the renewable vitality, space travel and the then-prospering internet.

On account of these thoughts, Musk connected, and was immediately acknowledged, into a doctorate program at Stanford University to concentrate connected material science. Two days into his time at Stanford, he exited the school to frame his first organization Zip2.

Chapter 3: Early Businesses

In the months preceding his making a trip to California, Musk viewed in interest as Netscape Communications opened up to the world and made a man more youthful than himself amazingly well off all the while. Now Musk's advantages totaled out at roughly $2,000 and an utilized auto yet he recognized what he needed to do and how to approach doing it.

He conveyed his sibling Kimbal out to California, obtained $28,000 from his dad and began a product organization named Zip2.

Zip2

Zip2 was a business that sold city directs that could then be utilized by daily papers as an approach to make themselves pertinent in the youngster online space.

Zip2 was made after Musk met the makers of Navteq, a computerized mapping organization and persuaded them to give him a chance to utilize their online maps.

He then bought an advanced index of organizations in his general vicinity, joined them utilizing a smidgen of code and made one of the principal computerized mapping administrations. Furnished with his inventive item, Musk soon ended up pursuing various significant daily papers and not long after that, offering his administration by means of their sites.

Notwithstanding their initial achievement, the Zip2 workplaces were in an once-over office fabricating that additionally served as the Musk siblings' essential place of habitation. The roof spilled exorbitantly and the main furniture to discuss was a couple of futons and a work area for Musk's PC which additionally served as the essential server for the greater part of Zip2's movement.

Amid the nights Musk would utilize the program to code and amid the day it would associate with the web by means of an opening in the floor Musk penetrated to exploit the online association accessible in the business first floor.

As Zip2 developed, it pulled in the consideration of financial specialists and in 1996 a funding firm offered the organization $3 million, however Musk needed to step aside as the CEO and let a business official named Richard Sorkin assume control.

The investment firm felt Sorkin was a superior decision since he had a degree from Stanford. At first it appeared as if everything would have been smooth cruising as Sorkin immediately landed contracts with real daily papers, for example, the Chicago Sun and also the New York Times. Tragically for Musk, Sorkin then started to search out extra speculation pay from a few of these expansive daily paper chains too.

Musk didn't care for going up against financial specialists from the client base the organization should target however wound up with little to do as when he understood what was going on he just possessed 7 percent of the majority of Zip2's stock.

All things considered, he was stuck viewing different new companies, for example, Yahoo! moving towards the future while his organization was profoundly settled in the methods for old media. In spite of establishing the organization, he was currently stuck in a VP part with little to do yet watch his organization's significance disappear.

Later that same year, Musk got word that Sorkin was chipping away at an arrangement that would offer Zip2 to an internet searcher named CitySearch which would have made an across the nation form of the nearby web search tool.

Musk chose this was the place he needed to take a stand or be left with nothing to battle for. He spread expression of his dissension inside the organization and arranged a revolt with an end goal to get Sorkin expelled from power.

With a hefty portion of the greatest names in the organization behind him, Musk figured out how to effectively get Sorkin expelled from the organization. This is the place his prosperity obstructed, be that as it may, as opposed to restoring him as CEO as had been his arrangement, the barricade was still made basically of Newspapers and different individuals from the old monitor who rather put Mohr Davidow's Derek Proudian in control.

In that capacity, while the CitySearch arrangement was pulverized, Zip2 was soon sold to Compaq rather who paid $300 million, the most a web organization had ever been esteemed at, and collapsed it into its aggregate with no place in the bundle for Musk.

At the time, Musk was disparaging of the deal taking note of that the investment firm ought to have abandoned him in control as "Incredible things never happen with expert chiefs or VCs [venture entrepreneur firms] in control, they don't have the knowledge or the imagination."

At its pinnacle Zip2 served 120 daily paper bunches countrywide, and its definitive deal netted him an individual $22 million. In spite of this, Musk has constantly considered Zip2 a disappointment since he had wanted to shape the web all the more effectively, with the advantage of the normal individual his essential concern.

The way that he rather made a route for a diminishing industry to stay important for a modest bunch of extra years stayed with him which is the reason his next wander was composed from the beginning to challenge the norm every step of the way.

X.com

In 1999, Musk was prepared to pitch his next thought, a money related administrations stage which worked without the requirement for customary banks, to Sequoia Capital an outstanding firm that any semblance of Oracle, Cisco and Apple had all used to get their begin.

That same day he left the meeting with $25 million in speculation capital for his next wander, a site known as X.com. While the first thought that Musk pitched to Sequoia included many if not the majority of the components that PayPal offers today, the adaptation of X.com that propelled in 1999 was altogether downsized and rather hyper-concentrated on making individual to individual online installments a reality.

The organization saw early achievement and he was soon offered an alluring arrangement from an organization called Confinity. Confinity needed to converge with X.com yet leave Musk in control as the CEO of the new organization. All the more vitally, it would give Musk access to Confinity's PayPal programming which was specifically contending with X.com.

While the innovation that every organization conveyed to the table mixed great, the same couldn't be said of the people that chipped away at the beforehand isolate X.com and PayPal groups.

The threatening vibe and micromanaging implied that Musk spent the better part of a year contending with different identities, dreams and inner selves. This wasn't what Confinity originators Peter Thiel and Max Levchin were searching for when they consented to give Musk a chance to be CEO, in any case, and the following year when Musk left the nation to meet with new financial specialists Levchin and Thiel utilized his nonattendance to arrange him from his position as CEO.

The control of the organization returned to the Confinity match who changed the name of the organization to PayPal and sold it to eBay the following year for $1.5 billion in real money and shares of eBay stock. While no more drawn out in control of yet another organization, Musk still held 11.7 percent of PayPal's aggregate stock at this crossroads which implied the deal left him with $160 million.

Chapter 4: Current Projects

While he positively has various different pots in the fire, Musk is at present basically centered around running two organizations, SpaceX and Tesla Motors and both are effectively changing assumptions about their particular businesses.

SpaceX

Musk has never been bashful about imparting his insights on space travel, which he accepts is essential to the long haul survival of mankind. He is likewise attached to stating that he trusts he will kick the bucket on Mars, however not amid the underlying landing. In view of that, is it any ponder that he invested the energy after he was evacuated as CEO without bounds PayPal yet before getting his PayPal buyout payout, chipping away at what might in the long run get to be distinctly known as the Mars Oasis extend?

The objective of the venture was to dispatch a little nursery into space and control it remotely until it arrived on Mars, prepared to work and loaded with everything a space traveler would need to convey a touch of green to the red planet.

Thus, Musk additionally planned to restore enthusiasm for space go to another era which he accepts is urgent as it may be, measurably, simply an issue of time before an annihilation level occasion makes earth a great deal less tenable that it at present is.

Beginning in 2001, Musk started fabricating an association with merchants of ballistic rockets in Russia. His underlying welcome was frosty, be that as it may, as the Russians felt he didn't recognize what he was discussing.

Amid his next excursion to Russia in 2002, Musk demonstrated he was presently a specialist on the point which is the reason his new companions offered to offer him one previously owned ballistic rocket for $8 million.

In the wake of running the numbers, Musk went to an acknowledgment that found him totally napping, it would be less expensive for him to begin an organization and construct his own rockets than it is purchase a solitary Russian rocket.

When he started to run the numbers much more completely, Musk additionally understood that the overall revenue on space faring vessels was cosmic, with just somewhat more than 1 percent of the deal cost being sufficient to take care of the whole expense of development.

Also, with an organization whose express reason for existing is to assemble and offer rockets, it would be much simpler for him to plan a ship prepared to do effectively making it the distance to Mars.

With the numbers supporting the choice, Musk soon framed Space Exploration Technologies (SpaceX) and, working with the group he had amassed, soon created plans for a vehicle equipped for achieving Mars effectively while as yet being less expensive than whatever other rocket available by a cosmic 90 percent!

Besides, SpaceX configuration still returns a normal of 70 percent benefit of the speculation cost per rocket. To get SpaceX up and running, Musk spent generally $90 million of his own cash on the venture just in light of the fact that he thought it was the best thing to do.

SpaceX's witticism is to make mankind a spacefaring race and they are as of now doing their part to make this fantasy a reality. Before concocting the Mars Oasis extend, Musk had never to such an extent as taken a class in fly impetus, a great deal less advanced science.

Unwilling to let something like an absence of any earlier information stop him, he got to be companions with a neighborhood scientific genius and requested that acquire his exploration on the point. At the point when the scientific genius asked which inquire about, Musk essentially motioned to the closest bookshelf brimming with books on the theme. He then took them home and read them all.

At the point when this didn't lead him to the level of information he was seeking after, he then went out and contracted each and every accessible scientific genius and picked their brains until he was agreeable he knew everything there was to think about the subject.

Keep in mind, this was the level of research and commitment that Musk put into the subject when he was still essentially anticipating purchasing a Russian rocket. By the by, amid this stage he was at that point conceptualizing thoughts of his own and counseling with his scientific geniuses on the arrangement that would at last prompt to Falcon 1, the principal rocket transport his group manufactured which he named to pay tribute to the Millennium Falcon.

Extra wellsprings of motivation for SpaceX originated from the Foundation arrangement composed by Isaac Asimov. Musk has as often as possible expressed that Asimov's perspectives with respect to the right utilization of space travel advancing is a critical stride with regards to extending the human awareness past its as of now generally restricted extension.

At present, Musk says that mankind has taken in regard to as long to create space venture out as it did to slither out of the seas a great many years back which implies we are just about due for a noteworthy change.

Since its beginning, SpaceX has created a couple of unmanned dispatch vehicles, Falcon 1 and Falcon 9, and in addition a completely working unmanned shuttle dedicated Dragon.

Falcon 1 was the primary vehicle that had even made it to circle utilizing fluid fills and propelled by a privately owned business when it had its first flight in 2009. In 2012, NASA marked an agreement with SpaceX to put Dragon exclusively accountable for refueling the worldwide space station and also conveying the space travelers supplies from the surface.

Dragon is currently utilized rather than the real space carry for space flights which implies Musk reexamined space go in under 20 years. Off by a long shot to fulfilled, SpaceX propelled its first satellite into space in 2013 and in 2015, the Falcon 9 propelled a more propelled satellite worked to watch the atmosphere of profound space and decide exactly how precisely sun oriented flares influence electromagnetic fields on earth.

For the following stride in this procedure, SpaceX is at present attempting to get authorization from the United States government and extra governments worldwide to dispatch a large group of 4,000 satellites which would then be utilized to guarantee everybody had admittance to solid and quick web associations.

Furthermore, SpaceX can as of now be said to be the most productive makers of rocket motors in the whole world and their Merlin 1D display motor, which is sufficiently capable to lift more than 40 autos, is as of now utilized for various purposes far and wide.

SpaceX was given its first NASA contract in 2006 alongside almost 2 billion dollars to get the Merlin 1A working appropriately. These tests eventually prompted to the making of the Falcon and the Dragon.

With SpaceX, Musk wants to diminish the cost of space go to the point of making a sensible idea once more. He will likely send a kept an eye on mission to damages by 2030 and a settlement of almost 100,000 by 2040.

He had been quotes as saying, shockingly, on the off chance that he has anything to say in regard to it, everything on Mars will be all electric. Keeping that in mind, he has additionally made the Musk Foundation with the trust of deciding the best renewable and clean vitality sources can be sued to make space travel speedier, less expensive, more secure and more proficient.

Tesla Motors

While Musk was caught up with agonizing over rockets, various different specialists were chipping away at the model for what might in the end be known as the Tesla Roadster. Having been considering electric autos for the majority of 2 decades, in any case, Musk went ahead board when the venture asked for subsidizing.

He put vigorously in the venture and went up against the part of executive of the load up around then too. This position additionally permitted him to go up against more dynamic part in the outline of the forthcoming monetarily discharged Teasla Roadster.

At present, Tesla Motors offers 3 distinct models out and about with a fourth form, estimated around the cost of whatever other comparative vehicle, declared in March of 2016 which rapidly sold through its preorder distribution. While the quantity of vehicles it has out and about is still generally little, Tesla Motors is as of now being contrasted with the Ford Company.

It is additionally the primary fruitful new car organization to be established in America in over a century. While he was unquestionably required in the meeting up of the Roadster's definitive outline, Musk did not start to go up against a more dynamic part of dealing with the organization until 2008, around the begin of the Global Financial Crisis.

He went up against the part of CEO around then and also the part of item draftsman too. While the product offering has since turned into a win, it was a long way from a beyond any doubt thing at a few focuses in the generation procedure. Musk initially met Marc Tarpenning and Martin Eberhard, the makers of the first Tesla show in 2001 when the pair went to listen to Musk talk on his most up to date energy venture of setting out to Mars.

They traded merriments and not a lot happened to the experience with the exception of that Musk recalled their names when they met again three years after the fact in 2004 to pitch Musk on the thought for an electric auto called TZero. Musk was in a split second snared on the pitch and organized to meet with the match the exact one week from now.

The meeting, booked for a tight 30 minutes, rapidly developed to be over 3 hours long as the trio talked about the specifics of the TZero and also the significance of making a vehicle that could contend out and about and also at the pump.

This meeting was additionally the beginning of the rollout system that Tesla Motors would at last utilize to awesome achievement, beginning with a top of the line model to catch the hearts and psyches of general society before revealing a standard release to exploit the intensity.

This early meeting to generate new ideas likewise required the primary vehicle in their stable, the Tesla Roadster, to move of the line, with the suspicion that the organization would be effortlessly turning a benefit by 2008.

This eventually turned out to be essentially excessively idealistic, be that as it may, as the coming weeks and months would see Eberhard and Musk butting heads regularly while the formation of the vehicle itself grieved in the midst of various creation and designing issues.

It was amid this period that Musk likewise played a more involved part in the production of the organization's lead vehicle, rolling out a few improvements including the entryway situation and the choice to make novel headlights, setting things back considerably.

Additionally postpones gathered as he saw the need to overhaul the seats, the style of the inside and even upgraded the transmission from the beginning. While the progressions were made to upgrade the general look, feel and nature of the vehicle to guarantee it was justified regardless of its exceptional value, they pushed the officially tight generation plan more distant than it could permit.

The choice to utilize one of a kind parts likewise put the juvenile organization into a positon of sourcing the formation of one of a kind parts, a procedure nobody working at the organization at the time was really acquainted with.

This prompt to Musk boring significantly more duty, going so far as to visit Lotus, the assembling organization in England that would at last make every Roadster just to attempt and understand the spiraling generation line.

In spite of these hardships, Musk clutched his craving for the Roadster to be recognizable, and available and it wound up being the essential imaginative compel behind a definitive outcome.

In 2007, in the wake of battling with Eberhard almost consistently, an especially warmed contention where Musk was credited with the formation of the Roadster in an article and Eberhard was not said, Eberhard left the organization in the midst of a whirlwind of claims and countersuits by Musk.

Now Musk went up against the obligations of CEO, making his first demonstration in the position to flame a fourth of the current staff as the extended creation plan implied the organization was discharging cash and it was the best way to guarantee his $50 million venture wouldn't turn into a tremendous oversight.

In 2008, the primary Tesla Roadsters at last moved off of the line and the surveys were ordinary, best case scenario and shocking even under the least favorable conditions. By 2010, 75 percent of the underlying run had been reviewed because of different equipment and programming mistakes. The underlying keep running of the Roadster finished with 2,150 units being created and delivered to more than 30 nations around the world.

Not to be hindered by what he saw as only a more extensive field test for the idea, Musk multiplied down on his Tesla Motors venture and started taking preorders for the follow up to the Roadster called Model S. Taking after the influx of positive early criticism with respect to the Model S, the organization documented an IPO worth $100 million, generally twice as much as Musk had sunk into the organization up to this point.

Toward the beginning of 2016, the organization was assessed to be worth $25 billion. This still isn't sufficient for the present CEO who says he anticipates that the organization will be worth almost 30 times as much before 2030.

After the Model S propelled to altogether enhanced surveys and began taking off in the way that Musk and alternate makers had imagined each one of those prior years, Tesla Motors presented a 4-entryway variety of the Model S and their first electric games utility vehicle, the Model X.

The organization has additionally started to fabricate the powertrain framework that drives the present electric offerings from Toyota and also Mercedes.

When he assumed control as CEO, Musk saw the most serious issue identified with the across the board selection of the electric auto to be its restrictions with regard to the capacity to travel a broadened remove, paying little heed to the span of an individual battery.

On account of this he started an activity to expand the quantity of charging stations accessible over the United States. In the previous 8 years this activity has dramatically multiplied the accessibility of charging stations across the nation. In 1992, a couple of Musk's cousins requested his considerations on venture openings in California.

Musk pointed them towards sun powered power and in 1993 SolarCity was conceived. In 2007 it turned into the biggest single provider of sun based power in California. Musk is additionally director of the board on this organization and it was a substantial part of the charging station activity both in California and the nation over.

Musk's objective all through his residency as the CEO of Tesla Motors has been to expand the across the board acknowledgment and use of electric vehicles when all is said in done, not simply of those flying the Tesla Motors signal.

To this end, he has likewise discharged the greater part of the licenses identified with electric engine innovation that the organization already held. This implies anybody is welcome to utilize their outlines the length of it is benefited in confidence, with the goal of expanding the general improvement of the item space.

At present Musk's compensation with the organization is only $1 with whatever else being created from investment opportunities and execution rewards. In spite of having been in general society eye for the majority of 10 years, Tesla was just ready to secure the rights to Tesla.com in 2016.

Preceding this point, the area was held by a man named Stuart Grossman who initially acquired the space in 1996 preceding anybody had considered purchasing up clear areas for future benefit.

Grossman wasn't utilizing the space for anything, however he enjoyed having the capacity to sooner or later which is the reason it took an individual visit from Musk to at last persuade him to offer.

While a definitive cost of the exchange was not unveiled, a dear companion of Musk's named Jason Calacanis was cited afterward as saying it would have been justified regardless of a few million in Musk's eyes to secure the space being referred to.

Chapter 5: What's Next

As an undergrad, Musk asked himself what three advances would change the world the most in his lifetime and afterward went to deal with making his blemish on every one of them.

As individuals are just a couple of years from having the capacity to sensibly have one of his autos drive them around while they utilize PayPal to purchase something with assistance from web from his satellites, it is protected to state he hit his objectives.

This doesn't mean he is prepared to surrender it all, in any case, as he has extra arranges that guarantee the future will be the most ideal form of itself it can be.

Hyperloop

In 2013, Musk went to a presentation of the condition of California's recommendation for a rail framework that would go at high speeds between Los Angeles and San Francisco.

He didn't think the framework that many specialists had invested years creating was quite that quick so he chose to concoct something better. To do as such, he got the greater part of the architects from his electric auto organization and the greater part of his scientific geniuses together and worked with them to think of what they are calling the Hyperloop.

The Hyperloop is a method for voyaging long separations at rates of 700 miles for each hour inside containers that go on pads of air. The concluded plan of the Hyperloop, and in addition subordinate gadgets, took about a year to make and once it was done, Musk discharged the schematics on the web and pronounced them open hotspot for everybody to utilize.

He then declared an opposition, supported by SpaceX, to plan the best case for use in the Hyperloop framework. This opposition, at present in progress, pits groups from around the globe against each other, testing their true plans on an enormous test track that Tesla made for simply this reason.

The champs will get chances to further improve their outlines through SpaceX and will be picked in light of the functional use of their plans in late 2016.

Open AI

In 2015 Musk uncovered the presence of yet another activity, this one a charitable that is committed to exploring and making counterfeit consciousness in a way that guarantees it advances to be both sheltered and helpful for everybody included.

In particular, Musk has expressed that he needs the association to remain against any potential mishandle of the masses that could be executed by a counterfeit consciousness that is made for such a reason either by a noteworthy partnership or world government.

Musk isn't the only one in this attempt either, and has secured the union of other outstanding logical personalities including Stephen Hawking who trust that counterfeit consciousness may well represent the most true threat with regard to its capability to contrarily effect mankind's survival rates long haul.

This is the reason Open AI exists, to guarantee that counterfeit consciousness benefits humanity as opposed to decimating it. Like the vast majority of the innovative achievements he has been connected with, the greater part of the work that Open AI wills be open-source and unreservedly accessible on the web.

Musk is the co-seat of the venture and is exceptionally mindful that the association needs to tread deliberately to guarantee they don't coincidentally make the thing that they fear the most.

He trusts that the most ideal approach to get ready against what could be viewed as the darkest course of events, is to set up the masses for what may be sooner rather than later.

Musk plans to give enough promptly accessible data and programing that everybody has the instruments to secure themselves if, or when, the time comes.

While Musk trusts that the not-for-profit will in the long run make something that outperforms the knowledge of its makers, he doesn't anticipate that that will happen at any point in the near future, not for a long time at any rate.

The venture is subsidized and prepared for the test, be that as it may, as people from around the globe have officially promised over $1 billion to guarantee the association's coffers are full for quite a long time to come.

Manmade brainpower is the one vision that Musk has not yet acknowledged and this time there is not really anybody anyplace who will wager against him.

Chapter 6: Sun based City

SolarCity partnership is an American vitality administrations supplier, with its central station situated in California. Its essential administrations are the outline, deal, financing, and establishment of sun powered and supportable power frameworks all through the United States.

SolarCity is likewise led by none other than Elon Musk. Musk plainly has an extensive enthusiasm for the maintainability of the Earth and its occupants, as confirm from his putting resources into tasks, for example, Tesla and SpaceX. SolarCity was initially established in 2006 by Lyndon Rive and Peter Rive – Elon Musk's cousins.

Elon is currently the biggest shareholder of SolarCity and has even proposed a 2.6-billion-dollar bargain that would see his Tesla organization buy SolarCity. Musk is imagining the marriage of the two organizations, which would see Tesla engine vehicles be fitted with an altogether new rooftop, totally made up of sun powered boards.

His vision is to make "the world's lone vertically incorporated vitality organization offering end-to-end clean vitality items". The client would have the capacity to have a SolarCity board introduced on their rooftop, catch the vitality from the sun and store it in a Tesla battery, and after that utilization that to control their home or auto.

This would permit individuals to basically live off of the network with the help of only one organization! Regardless of whether this arrangement will experience still stays to be seen in any case, as the choice descends to Tesla's board individuals.

There is a touch of wavering encompassing the arrangement, as made by shareholder weights. A few shareholders furthermore individuals in the media are review this proposed buyout as an awful arrangement. SolarCity is at present not benefitting, and has encountered an extensive drop in shares – making its present market esteem a ton lower than the proposed 2.3 billion-dollar price tag.

The arrangement clearly has its dangers, yet Elon Musk is no more interesting to hazard and has approached chapter 11 preceding with Tesla Motors. In any case, with enormous hazard comes huge reward. In the event that the interest for sun based vitality keeps on ascending, over the coming decades Elon Musk, Tesla, and SolarCity together would be in an awesome position to take the lion's share of the market.

Will the arrangement work out as intended? Be that as it may, in any case, Elon Musk is making huge moves, and will keep on making them, in the territory of sun based and practical vitality generation and conveyance.

Chapter 7: Elon's Investments

Over his profession, Elon Musk has made a wide range of speculations. In this section we will share the distinctive organizations he has put resources into throughout the years, both the effective ones, and the disappointments!

Zip2

Musk's first organization was Zip2. We have as of now examined this organization in a prior part, however basically Zip2 was the web's first business catalog, and could be considered a significant enormous achievement. Musk sold the organization in 1999 to Compaq for $307 million. At the time, that was the biggest sum ever paid for an online organization.

X.Com

Likewise beforehand talked about, Musk took $10 million from the offer of Zip2 and utilized it to begin X.com, an online installment benefit. In the long run, X.Com converged with a contender called Confinity, which later changed names to Paypal. Musk had a few issues with his new business accomplices and was in the long run voted off of the board.   It wasn't all terrible news for Elon be that as it may, as Paypal was obtained by eBay in 2002 for $1.5 billion. Musk was not a larger part proprietor by any methods, but rather regardless he netted over $100 million from the deal – an extraordinary profit for his underlying $10 million venture.

Everdream Corporation

Elon Musk's cousin, Lyndon Rive was a fellow benefactor of the organization Everdream. Everdream sold desktop administration administrations to private companies, and performed undertakings, for example, settling antivirus programming, and going down information.

Musk put resources into Everdream in the fourth round while he was still required with Paypal. Very nearly 10 years after the fact, Everdream was sold to Dell in 2007. SpaceX   In 2002, Elon Musk established Space Exploration Technologies, also called SpaceX. Elon trusts that this venture is a standout amongst the most imperative things to happen ever.

Clearly, it is his most prized organization and the one which implies the most. The objectives of SpaceX are to make rockets more reasonable, furthermore to permit people to wind up distinctly a between planetary species. SpaceX has had it's share of skeptics, naysayers, and monetary issues.

Be that as it may, at this present minute in time the future searches very splendid for SpaceX! In April 2016 SpaceX made gigantic ground by propelling and effectively re-finding a Falcon rocket.

In the coming years, Musk intends to send people in his rockets to Mars, to start colonization of the planet. The Musk Foundation   In 2002 Elon established the Musk establishment close by his sibling Kimbal.

The establishment grants gives that bolster look into maintainable and renewable vitality, space investigation, instruction, and youth sicknesses and disarranges. Tesla Motors   Musk put resources into Tesla Motors in 2004. The organization was initially established in 2003 by Martin Eberhard and Marc Tarpenning.

After his speculation, Musk joined the directorate. From that point forward, Elon has been intensely required with Tesla Motors with the point of making electronic autos the method for what's to come. After the market crash in 2008, Elon turned into the CEO of Tesla, a position which despite everything he holds today.

Tesla has had a few money related issues throughout the years, and has required a tremendous measure of speculation. Be that as it may, the organization as of late broke records by pre-offering its model S vehicle, totaling more than 325,000 deals in a solitary week! The model S car begins at $35,000 and is relied upon to start creation toward the end of 2017.

Surrey Satellite Technology

In 2005 Musk acquired a little 10% stake in Surrey Satellite innovation, a little satellite supplier. The thought behind the venture was not such a great amount of concentrated on making an arrival, yet rather for SpaceX to pick up a superior comprehension of how they would one be able to day work close by this organization in giving little and cheap rocket.

SolarCity

Already specified in this book, Musk has put into SolarCity. Clearly being extremely enthusiastic about reasonable and green innovation, this organization is a conspicuous venture for Elon. At present, Elon Musk is going about as the executive of the organization.

Mahalo.com

Mahalo.com is a question and answer site that was initially established in 2007. The site basically permits individuals to ask and answer questions. Be that as it may, when Google changed it's calculation in 2011, Mahalo encountered an expansive drop in business and was compelled to lay off 10% of its staff. The organization has since changed its system, and now principally concentrates on the most proficient method to recordings and live question and answer sessions.

Stripe

Stripe propelled in 2010 and is Paypals principle contender. After his involvement with Paypal and X.com, Elon Musk considered this to be a conspicuous venture to make. The organization as of now gives installment preparing administrations to online applications and organizations, for example, Lyft, Facebook, and Twitter. Stripe has encountered quick development and was as of late esteemed at $5 billion. This one was unquestionably a decent venture for Elon Musk.

Halcyon Molecular

Halcyon Molecular was established in 2008, with the objective of opening the greatest privileged insights covered up in DNA. The organization had vast objectives, and meant to give a human genome sequencing administration that cost under $100. Be that as it may, because of solid rivalry it wasn't intended to be. In 2012 the organization shut its entryways, getting to be distinctly one of Elon's initially fizzled ventures.

Tesla Science Center

Less a speculation but rather more it was a gift, Musk gave $1 million in subsidizing for the advancement of another science exhibition hall. The gallery is named the Tesla Science Center and is situated in New York. The gallery, named after the colossal researcher and innovator Nikola Tesla, was made in 2014. Elon Musk arrangements to construct the world's snappiest Tesla charging station in the exhibition hall's parking garage!

Vicarious While Musk is not a devotee of counterfeit consciousness, he put resources into AI organization Vicarious in 2014. The organization has been bizarrely shrouded with its undertakings, yet has expressed that they will likely make a PC that has a similar outlook as a human, however does not have to rest or eat. Elon thinking behind putting resources into AI organizations is to guarantee that AI is moving in the correct bearing, and is not being utilized or made for dangerous purposes.

DeepMind Technologies

DeepMind Technologies is another AI organization that Elon Musk has put resources into to. The organization was at first established in 2011, however was procured by Google in 2014, changing its name to 'Google DeepMind'.

The organization intends to join machine learning and neuroscience into an intense, universally handy, PC calculation.

Eventual fate of Life Institute

At the end of the day, all the more a gift than a speculation, Elon Musk gave $10 million to the Future of Life Institute in 2015. The establishment underpins inquire about into the dangers connected with creating innovations, for example, AI. This further shows how stressed Elon Musk is over the likelihood of manmade brainpower being utilized for negative purposes.

NeuroVigil

NeuroVigil is a startup that was propelled in 2007. The organization built up the worl'd first versatile mind screen, the iBrain. Musk turned into a central speculator in the organization in 2015. The organization utilizes its innovation to medication organizations lead clinical trials, and in addition to determine and treat patients to have neurological sickness.

NeuroVigil additionally needs to help NASA monitor their space traveler's brains while they're on board the International Space Station. This could be a major explanation behind Elon's enthusiasm for the organization.

Hyperloop

In spite of the fact that he hasn't physically put resources into the business yet, Hyperloop appears to be Musk's next huge venture. He initially declared the thought for Hyperloop in 2013, through which he needs to fabricate a progression of rapid rail frameworks that would transport individuals at paces surpassing 500mph! There is very little data about the proposed Hyperloop at the present time, yet stay tuned – the conceivable outcomes for this innovation look be enormous!

OpenAI

Another counterfeit consciousness organization, OpenAI, was initially established in 2015. Be that as it may, this time, Musk was one of the fellow benefactors. This organizations essential concentration is to end up distinctly a storehouse of research papers, blog entries, code, and licenses that driving researchers can contribute. OpenAI means to permit individuals to securely and transparently take a shot at AI. Elon Musk as of late contributed the $1 billion dollars of financing that the organization has raised in this way.

Chapter 8: Individual Life

Elon Musk has likewise had a quite fascinating individual life consistently. The weight of being always in the media, close by with money related battles and worry from work, can put a considerable measure of strain on a man's close to home life at home. Purportedly working 100+ hours weeks, it is straightforward how the man's family life can take a hit.

Elon was at first hitched to Justine Wilson for a long time, a relationship that created 5 children. Not long after their separation in 2008, he met Talulah Riley in a London dance club.

Musk was at the club, attempting to clear his head after the late separation. Presented by the club's promoter, Elon was in a split second inspired by the on-screen character, and in this way started their relationship. Elon was inspired by Talulah's knowledge and enthusiasm for discussing rockets and electric autos. He immediately understood that she was not just simply one more model, but rather had insightfulness to match her looks.

They got together again a few weeks after the fact in Beverly Hills, where lying in bed, Elon suddenly requesting that her wed him. He didn't have a wedding band, however later got her 3: a goliath one, and regular ring, and another with a jewel encompassed by 10 sapphires.

At the time in any case, his separation was not settled with his first spouse. He was additionally encountering money related issues with SpaceX and Tesla, and it was an exceptionally unpleasant time for the new couple. Talulah depicted him amid this period by saying, 'He looked like demise itself'.

Among the greater part of the anxiety, the couple separated, re-wedded not exactly a year later, and practically separated for a moment time. They altered their opinions and Musk tore up the legal documents before their second separation was handled. Elon gives off an impression of being an intense man to live with, experiencing just about 3 separates in a brief timeframe.

His first spouse, Justine depicted him by saying "Elon does what he needs and he is tenacious about it. It's Elon's reality and whatever remains of us live in it". A decided and centered man, Elon Musk is the result of an intense childhood in South Africa, where it is said he encountered mishandle from his dad.

A casualty of extreme harassing growing up, Elon has all the earmarks of being left with changeless passionate scars. Justine alluded to him as being "a tormented soul".

While his unpleasant up-bringing has formed him into an occasionally troublesome man to live with, it seems to have favored him with steadiness and core interest. He has a level of persistence that not very many others have, and maybe that is the thing that permits him to be the inconceivable business person that we know today.

Chapter 9: Counterfeit consciousness

As you may have seen from Elon Musk's diverse ventures, he holds a lot of enthusiasm for the zone of manmade brainpower. This intrigue however is coming for the most part from a position of dread and stress over what the eventual fate of manmade brainpower may hold. One of Musk's latest ventures, OpenAI, is centered around making the eventual fate of counterfeit consciousness protected and supportable.

There is an undeniable plausibility that counterfeit consciousness could soon outperform the insight and abilities of people. There are a few motion pictures portraying this conceivable end, for example, iRobot and Terminator, however a rendition of these anecdotal shows could eventuate.

Elon Musk has expressed that generally he isn't excessively stressed over computerized reasoning. He has putting resources into a wide range of AI organizations just to keep his finger on the beat of what's happening in that industry.

In any case, in a late meeting he said there is truly just a single organization that he's stressed over. He was hesitant to state which organization he was discussing, however emphatically indicated towards Google.

The other principle danger that Elon has said in the territory of AI is something many refer to as the Cyber Grand Challenge. The Cyber Grand Challenge is fundamentally a hacking competition held in Vegas. The Challenge is controlled by the Defense Advanced Research Projects Agency, additionally alluded to as DARPA.

DARPA expresses that their objective is "to discover new systems for countering digital fighting". Musk however isn't a major devotee to the legitimacy of this mission. He fears that the hacking test may come about inevitably in making something like the Skynet AI that existed in the Terminator motion pictures.

Fundamentally, he's stressed over it prompting to the making of an almighty supercomputer. DARPA looks to make a robotized computerized reasoning framework that can recognize and resolve bugs in a PC framework. Surrendering a PC over to this undertaking without the help and basic leadership forces of a real human could conceivably be lamentable.

For the vast majority, the likelihood of a vindictive manmade brainpower framework appears to be several years away. Be that as it may, with the present rate of mechanical improvement, it could happen a great deal sooner than many would anticipate. Fortunately, Elon Musk is exceptionally mindful of this issue and is utilizing his energy and impact to bring issues to light.

The eventual fate of computerized reasoning is generally greatly energizing and introduces some astonishing open doors for humankind. On the off chance that it can be produced safy, with Elon Musk's association obviously, then this is a range that we ought to have nothing to stress over!

Chapter 10: What Will Be Elon Musk's Legacy?

Things being what they are, by the day's end, what will Elon Musk be best known for? What will be this staggering man's legacy? Will it be his work in the territory of counterfeit consciousness? Will it be for his commitments to maintainable vitality through his Tesla autos, and work with SolarCity? On the other hand maybe it will be his part in possibly populating Mars!

Elon Musk has a scope of activities underway, from endeavoring to populate Mars, to making reasonable vitality for the world, to making a Hyperloop framework that could change the way individuals travel!

With such a large number of various tasks, it's difficult to know precisely what Musk will be known for. The tycoon business visionary is still just in his 40's. He has decades and many years of potential in front of him, so for all we know he will start another venture inside and out that will be the one we recall that him most for.

Be that as it may, in this present creator's feeling, Elon Musk will be best known as a designer of sorts. An intense business person who's work has actually changed the substance of the world. Like individuals, for example, Albert Einstein, Thomas Edison, Nikola Tesla, his work will be recalled worldwide for a considerable length of time and years after his life his over.

His legacy will incorporate a scope of creations, undertakings, charitable endeavors, and moving minutes where he could beat the chances stacked against him.

For any individual who has huge goals as a business visionary, creator, or in pretty much any commendable attempt, Elon Musk as of now is and will stay to be an inconceivable wellspring of inspiration and motivation that it can be conceivable.

Being a motivation for future eras might just be the most important piece of his legacy when it is altogether said and done.

Chapter 11: Lessons to Learn

In the event that you investigate the lives of incredible people who leave a permanent stamp on history, you can simply locate a couple takeaways or useful tidbits to live by.

It doesn't require much push to discover lessons worth copying in the life of Elon Musk, and giving them something to do on an individual level can make it less demanding to trust in yourself and to take after your fantasies both in the short and the long haul.

Know about the signs around you and act when fitting: Musk isn't a fruitful compel in a few exceedingly progressed and gainful ventures since he is additional keen, rich or even to a great degree fortunate - however he is each of the 3. He is effective in light of the fact that when he was in school he checked out the world as it was at the time and found the markers that pointed him in the correct bearing, first to the web, then to space and electric autos.

Anybody in a comparative circumstance in the meantime could have seen a similar thing however Musk was the person who saw the signs and set aside the opportunity to translate them and put their data to great utilize.

Musk saw what should have been done, and after that did all that he expected to keeping in mind the end goal to guarantee that he was in the correct place so that when the perfect time went along he was prepared for it. Besides, didn't let the way that completing was troublesome, deter him from exploiting the achievement he knew was headed. Being astute is helpful, as is being persevering or devoted to a cause; by and by, being every one of the three and seeing how to best apply every ability is the thing that truly makes a man fruitful.

Take everything in walk: When considering the achievement that Musk has had throughout the years, it is similarly imperative to recollect the misfortune he confronted first when he was almost pounded the life out of, again when he was removed from Zip2, again when he was expelled from X.com lastly the dramatization before the effective dispatch of the Tesla Model S.

In every single one of these circumstances he could have failed the hard street and discovered something less demanding without even batting an eye. Musk could have been self-taught and afterward quit working for life after both of his web organizations were at last sold off lastly, he could have left Tesla Motors as a stupendous fizzled explore after the Roadster and concentrated on his totally fruitful business of building rockets.

Rather, he utilized every difficulty to inspire himself forward to new and beforehand phenomenal statures. The lesson to gain from this is clear, never let occasions that seem awful at face esteem get you down, take each as a suggestion to take action to enhance yourself in new and fabulous ways.

Never quit enhancing: From the minute he thought tocreate the online guide database with nearby business data, Musk has been improving. At X.com he could have been upbeat concentrating on individual to individual email exchanges and never gone ahead to make PayPal in its present frame, or he could have just purchased a rocket from Russia and never shaped SpaceX.

He could have remained a basic financial specialist at Tesla or thought the proposed California High Speed Transit plan was sufficiently quick. Rather, he saw defects in frameworks that other individuals would have considered adequate and as opposed to changing what he looked like at these issues he chose to change the world to suit his needs.

The lessons to be gained from this are self-evident, in the event that you can't locate the correct specialty for yourself, make something new. In the event that you never quit advancing on your past victories, you will never quit being fruitful. Take after the more troublesome way and in the event that you drive forward you will discover achievement.

Work harder than the opposition: Musk is right now the acting CEO of a couple of organizations that are every worth more than $1 billion and he has a notoriety of being into micromanaging.

He is additionally executive or co-seat of various different sheets, leader of various foundations and is as of now judging the Hyperloop rivalry. Musk is known to work 100 hours for every week in light of the fact that, in his words, on the off chance that you work more than twice as hard as every other person you will complete 3 times as much every year. Diligent work and persistence are the genuine spine of each sort of achievement. Search out your motivation:

At age 18 Musk was at that point discussing electric autos and at 22 he was looking to the stars. SpaceX was framed with the goal of making humankind into a spacefaring race by reigniting the world's enthusiasm for space travel.

Musk doesn't have little dreams or make incomplete move, he is all in on the grounds that he trusts he is taking after the way he was intended to take after. In the event that you ever would like to accomplish even a tenth of what he has finished, you have to set aside the opportunity to have an inside and out discussion with yourself and figure out whether you are truly doing what you were intended to do.

When you locate your actual reason, the following stride is to remain absolutely determined until you have accomplished it.

Be adaptable: While you ought to have an arrangement to accomplish whatever it is your actual reason ends up being, it is critical to not seek after it if the majority of the outer confirmation focuses the other way. This is the means by which

Musk could react to the data that the rocket he needed to purchase from the Russians was way out of his value extend and effectively make SpaceX all the while.

He was additionally compelled to rotate when the Tesla Roadster's creation plan had spun fiercely wild to the point that he was in a position to practically lose his speculation. Rather he moved with the punches, did what should have been done, regardless of how troublesome it may have been, and endured until he discovered achievement.

Decide your own level of accomplishment: When Zip2 was sold for $300 million, it was the biggest arrangement of its kind that had ever been finished, a metric of achievement in practically anybody's book. Not for Musk, notwithstanding, as he was persuaded that the administration could have been far beyond what it at last got to be.

He saw disappointment where many individuals would have been glad to discover achievement and it drove him to significantly more noteworthy statures as a result of it. Like Musk, it is imperative to hear yourself out and not give other's dreams of achievement or disappointment a chance to shading your own impression of your activities.

Conclusion

Much obliged for making to the end of the book! Ideally, gaining from the life of one of the best personalities of an era has given you various bits of knowledge into how you can enhance your own particular life by finishing on the things that made Elon Musk a genuine progress.

On the off chance that impersonation is the purest sort of honeyed words, then you can unquestionably do more awful than complimenting the present day Henry Ford. The following stride is to quit perusing and to begin putting the lessons you have learned in the previous pages to work in your own life.

Keep in mind, the length of you exploit the signs your general surroundings is giving you truly can't turn out badly. Take after your own way, focus on your own prosperity and finish anything you set your psyche to. At last, on the off chance that you delighted in this book, please leave a survey on Amazon and let others realize what you think. It'd be incredibly refreshing!